Read for a Better World™

STARS

A First Look

PERCY LEED

Lerner Publications ◆ Minneapolis

Educator Toolbox

Reading books is a great way for kids to express what they're interested in. Before reading this title, ask the reader these questions:

> What do you think this book is about? Look at the cover for clues.
>
> What do you already know about stars?
>
> What do you want to learn about stars?

Let's Read Together

Encourage the reader to use the pictures to understand the text.

Point out when the reader successfully sounds out a word.

Praise the reader for recognizing sight words such as *the* and *are*.

TABLE OF CONTENTS

Stars

What are those bright lights in the sky?
They are stars.

Stars are big balls of gas. Stars give off light and heat.

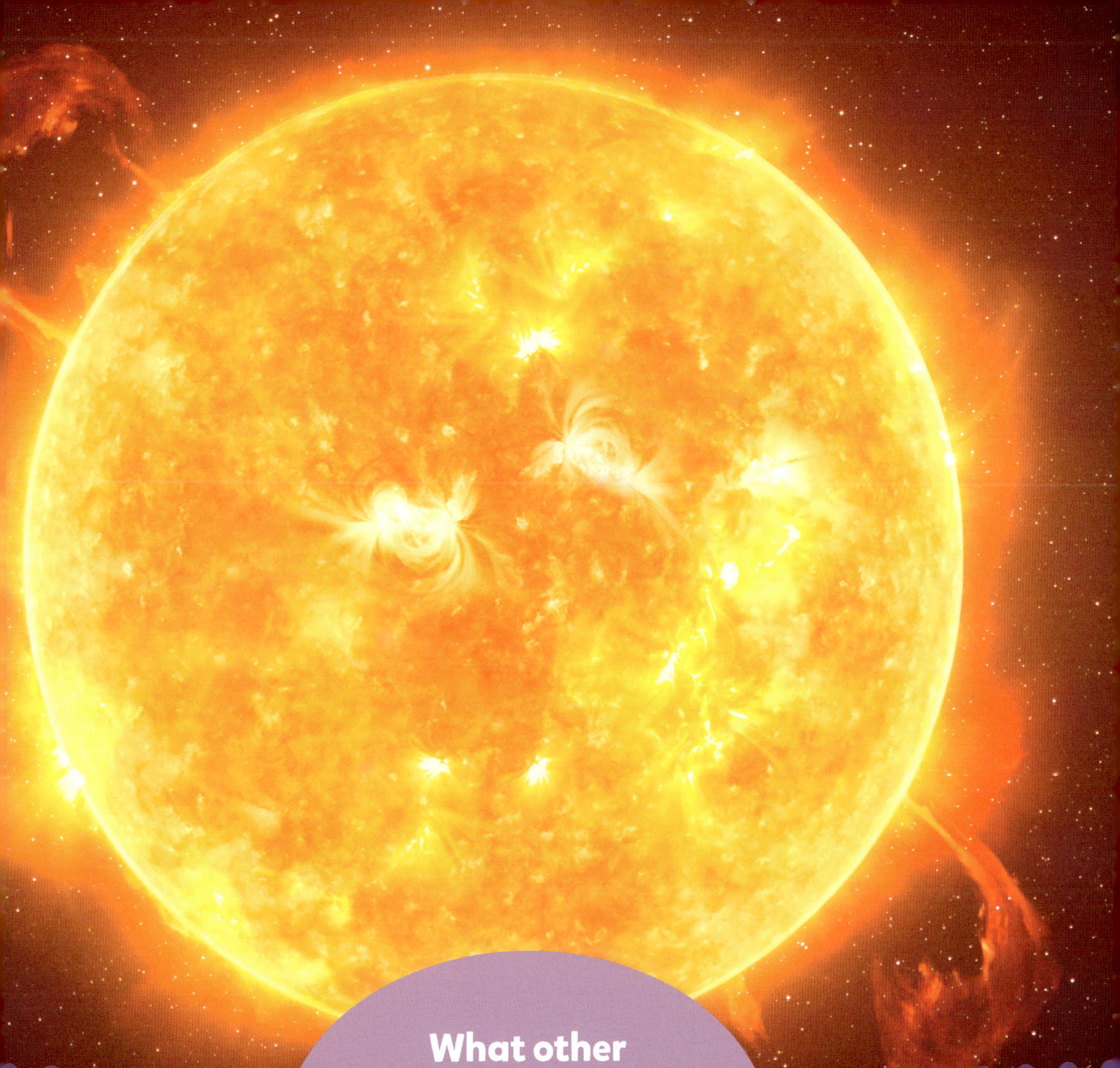

What other objects in space give off light?

There are trillions of stars in space. Stars are all different sizes.

Stars look small because they are far away.

Most stars are bigger than Earth.

Our closest star
is the sun.
The sun gives Earth
light and heat.

Stars do not
live forever.
When a star dies,
it explodes.

People give groups of stars names. This group is called the Big Dipper.

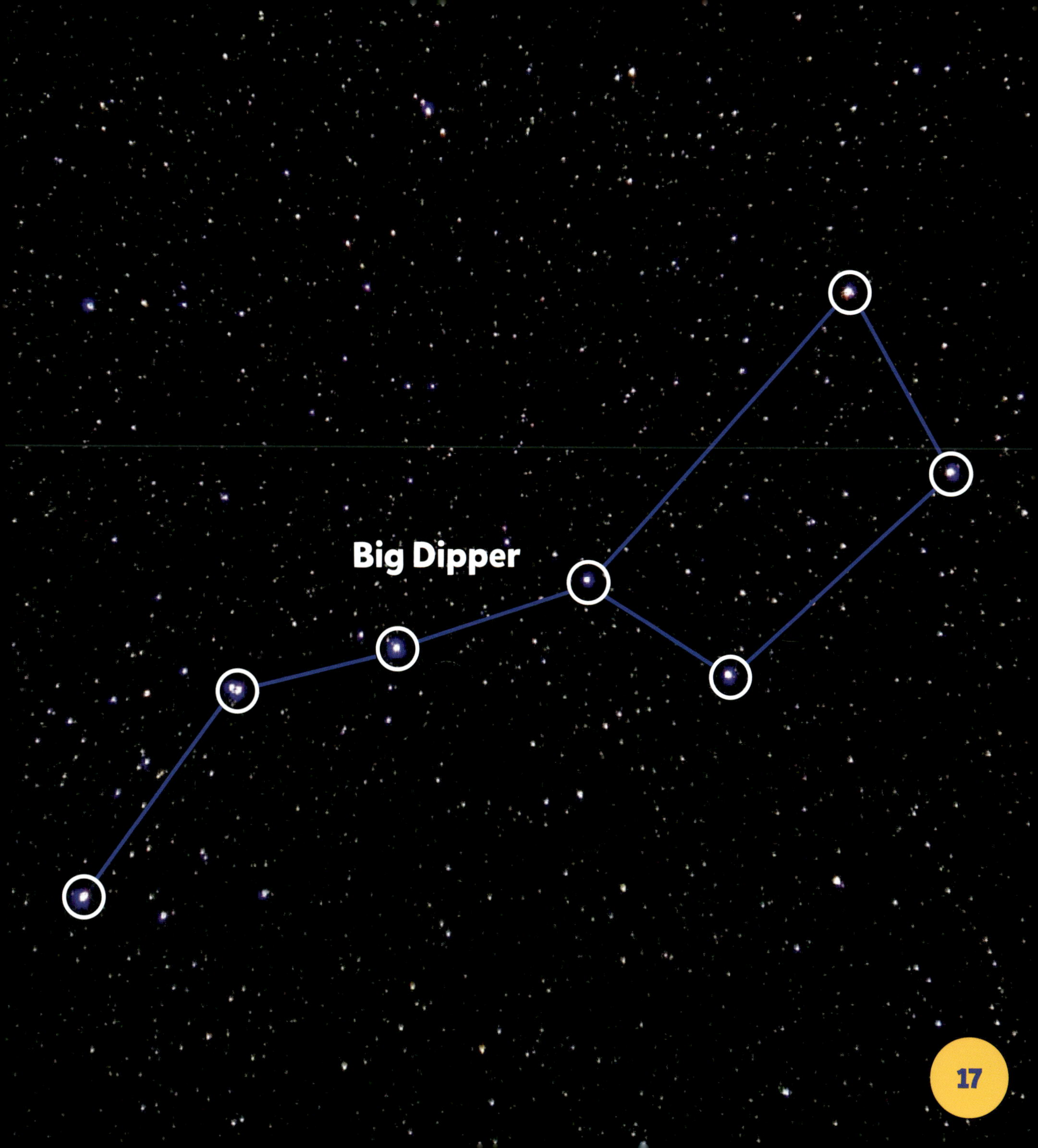
Big Dipper

We can see stars better with telescopes.

Have you ever used a telescope?

It is fun to look at stars.

You Connect!

What is something you like about stars?

Have you ever seen the Big Dipper?

Would you like to use a telescope?

STEM Snapshot

Encourage students to think and ask questions like scientists. Ask the reader:

What is something you learned about stars?

What is something you noticed about stars?

What is something you still want to learn about stars?

Photo Glossary

Big Dipper

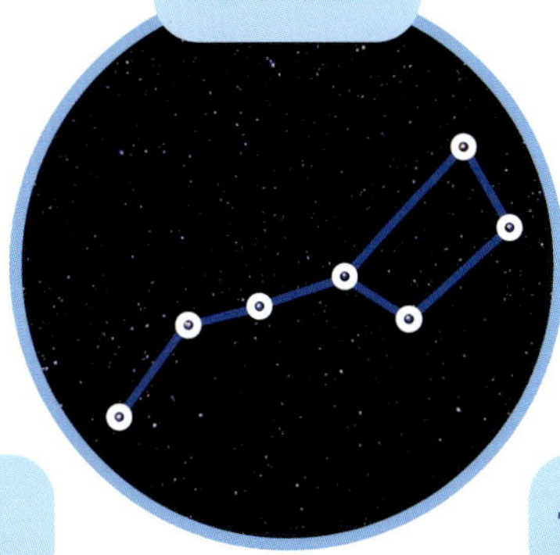

Earth

sun

telescope

Learn More

Drimmer, Stephanie Warren. *Night Sky*. Washington, DC: National Geographic, 2017.

Gardner, Jane P. *Constellations*. Minneapolis: Jump!, 2021.

Leed, Percy. *Sun: A First Look*. Minneapolis: Lerner Publications, 2023.

Index

Photo Acknowledgments

The images in this book are used with the permission of: © Abrill_/iStockphoto, pp. 11, 23 (Earth); © davidhajnal/iStockphoto, pp. 8–9; © designsstock/iStockphoto, pp. 12, 23 (sun); © dzika_mrowka/iStockphoto, pp. 6–7; © JaySi/iStockphoto, p. 10; © KIDSADA PHOTO/Shutterstock Images, p. 20; © miljko/iStockphoto, pp. 18–19, 23 (telescope); © NASA, pp. 16–17, 23 (Big Dipper); NASA Goddard, pp. 14–15, 23 (explode); © Smart-Photography/iStockphoto, pp. 4–5; © toxawww/iStockphoto, pp. 12–13.

Cover Photo: © Denis Belitsky/Shutterstock Images.

Design Elements: © Mighty Media, Inc.

Lerner Publications Company
An imprint of Lerner Publishing Group, Inc.
241 First Avenue North
Minneapolis, MN 55401 USA

For reading levels and more information, look up this title at www.lernerbooks.com.

Main body text set in Mikado a Medium.
Typeface provided by Hannes von Doehren.

Library of Congress Cataloging-in-Publication Data

Names: Leed, Percy, 1968- author.
Title: Stars : a first look / Percy Leed.
Description: Minneapolis, MN : Lerner Publications, [2023] | Series: Read for a better world | Includes bibliographical references and index. | Audience: Ages 5–8 | Audience: Grades K–1 | Summary: "Stars light up our night sky. Easy-to-read text and fascinating photos give readers a first look at these giant balls of gas"— Provided by publisher.
Identifiers: LCCN 2021051334 (print) | LCCN 2021051335 (ebook) | ISBN 9781728459240 (library binding) | ISBN 9781728464350 (paperback) | ISBN 9781728462202 (ebook)
Subjects: LCSH: Stars—Juvenile literature.
Classification: LCC QB801.7 .L44 2023 (print) | LCC QB801.7 (ebook) | DDC 523.8—dc23/eng/20211216

LC record available at https://lccn.loc.gov/2021051334
LC ebook record available at https://lccn.loc.gov/2021051335

Manufactured in the United States of America
1 - CG - 7/15/22